To St Nicolas
School
with love from.
Mrs Butterworth
2019

CT: For little Francesca.
RC: For my parents: John and Glenda Cleave.

BOUNCING BACK

A catalogue record for this book is available from the National Library of Australia.

Published by
CSIRO Publishing
Locked Bag 10
Clayton South VIC 3169
Australia

Telephone: + 61 3 9545 8400
Email: publishing.sales@csiro.au
Website: www.publish.csiro.au

Edited by Dr Kath Kovac
Cover design by Astred Hicks (Design Cherry)
Text design and layout by James Kelly
Printed in China by Toppan Leefung Printing Limited

The views expressed in this publication are those of the author and illustrator and do not necessarily represent those of, and should not be attributed to, the publisher or CSIRO.

Note for readers: Scientific terms are explained in the glossary at the end of the book.

BOUNCING BACK

AN EASTERN BARRED BANDICOOT STORY

ROHAN CLEAVE AND CORAL TULLOCH

CSIRO

PUBLISHING

I am an Eastern Barred Bandicoot.

My species is found only in Australia and nowhere else in the world.
We are also one of Australia's most threatened species, and are in
great danger of being lost forever.

I have a long, tapered nose, thin whiskers, sharp claws for digging,
a stripy pattern on my rump and a short tail.

Stripy pattern

Long tapered nose

Short tail

Sharp claws

Thin whiskers

Mother and babies

We usually only live for about two to three years.
This means that we grow and mature very quickly.

At just three months of age, females are old enough
to have their own babies.

We have one of the shortest pregnancies of any mammal –
just twelve-and-a-half days. When conditions are good,
we can have two to three babies at a time.

Our babies drink their mother's milk while snuggling
in a backwards-facing pouch. The pouch also protects us,
because we are only the size of a jellybean
when we are born.

We eat a wide variety of foods. These include moths, beetles, crickets, earthworms, grasshoppers and insect larvae.

Delicious!

We also eat some roots, grasses and seeds.

Getting all our fluids from the food we eat means
that we can survive without drinking water.

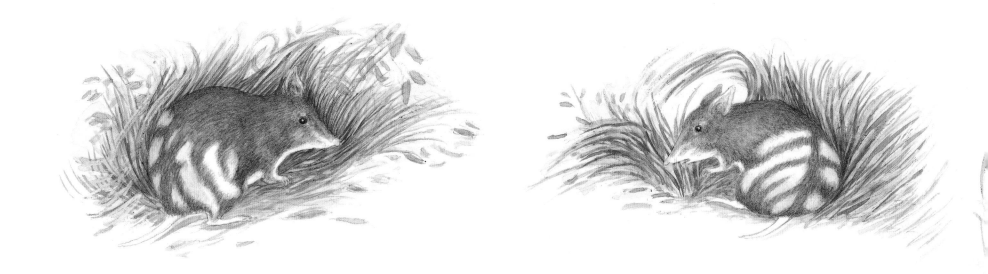

We build nests on the ground out of grasses, leaves and twigs.

For you they would be very hard to see, but for us they provide
the perfect shelter to protect us during the day.

We have a home range where we build a few different
nesting sites.

When night comes, we emerge from the shadows under the cover of darkness, exploring and feeding, hiding from our predators.

Eastern Barred Bandicoots were once common across Victoria and Tasmania.

But people started clearing the land of native grasses, and building roads and houses. Drought and bushfires destroyed our homes and left us exposed, with nowhere to hide.

Introduced animals, such as foxes, could hunt us easily. We became prey for owls at night.

We are still surviving in Tasmania. But in Victoria,
we have disappeared from the wild.

With so many threats and so few of us left,
we had to find somewhere new to live.

In Victoria, one of our last hideouts in the wild was at a
rubbish tip, where humans throw their garbage out of sight.

We found safety and protection in abandoned,
rusting cars that had also been dumped.

Once we fled to this unnatural environment, this place of
human rubbish and waste became the key to our survival.

The threat of falling prey to foxes was always present, but
living in the tip gave us some protection. The piles of garbage
and unwanted things became a place to hide and call home.

If we hadn't changed our ways and adapted, we could have
died out, and our species could have become extinct.

Even though we had somewhere new to live, we struggled
to survive in this strange place.

With our wild population still on the edge of extinction,
small teams of dedicated people banded together in
a desperate effort to save us.

They realised that urgent action was needed. If they didn't step in to save us, we may not have survived the constant threats.

People gave us safe places in zoos to have our babies and grow our population. Without these dedicated breeding programs, we could have vanished forever.

Captive breeding and research will help our species survive long into the future.

As well as growing our numbers, people decided that
the places where we used to live needed to be protected,
so that we could live there again.

They built fenced reserves on our former home ranges that
our predators could not enter, keeping us safe and secure.

The fences also helped to protect the environment.

The native grasses that we rely on to find food,
hide and build our nests in could grow once more.

Small, coastal islands are perfect places
for our recovery too.

Some of us were placed on these islands,
far away from foxes and other predators.

The islands provide a safe future for us.

Maremma dogs are also being trained to protect us.

For thousands of years, these clever and loyal dogs
have protected farm animals, standing guard
to help keep predators away.

Our story of survival is a tale that is still unfolding,
thanks to the many volunteers who have joined
the fight to save us.

Now, our future looks brighter.

Conservation of our little friends like the Eastern Barred Bandicoot
will help ensure their survival and allow future generations to
enjoy and connect with our native animals.

Every plant and animal has a connection to its natural ecosystem.
Small actions that you take can make a big difference to
conservation efforts by helping to restore these connections.

Let's make sure this beautiful little Australian marsupial survives
long into the future.

The Eastern Barred Bandicoot

Species name:

Perameles gunnii (mainland subspecies)

Perameles gunnii gunnii (Tasmanian subspecies)

The Eastern Barred Bandicoot is one of Australia's most threatened native species. It is a small, nocturnal, terrestrial marsupial that belongs to the Peramelidae family.

This book is about the mainland Eastern Barred Bandicoot, which used to be found in the wild from Melbourne through to south-western Victoria, and even into small areas of South Australia. The mainland subspecies was listed as extinct in the wild in 2013, but a second subspecies can still be found living in Tasmania.

Eastern Barred Bandicoots are around 30 centimetres long with a short tail of around 10 centimetres. Adults usually weigh 600–800 grams, but the biggest males can reach one kilogram or more. They have three to four very distinct, stripy bands on their rump. They dig small, shallow, conical holes to forage for food.

In the wild, the bandicoot's lifespan is two to three years. Females can reproduce from just three months of age and can breed throughout the year, but tend to have a break when the weather is hot. The young remain in the mother's pouch for 55 days before becoming independent.

The natural habitat of Eastern Barred Bandicoots is grassy woodlands with native, perennial tussock grasses. Since European settlement, the bandicoots have lost significant amounts of habitat over their home range due to human disturbance, land clearing and stock grazing. Bushfires and extended drought have also reduced suitable vegetation for nesting sites. When bandicoots are out in the open, they are easily prone to predation, especially from their primary threat – the introduced red fox.

The last wild mainland Eastern Barred Bandicoot population was recorded in Hamilton, in south-western Victoria. They were living at the local sports field, on the banks of a creek known as the Grange Burn, and most surprisingly, in the Hamilton tip. The tip provided protection from predation, allowing the bandicoots to hide among the waste and old car bodies. Unfortunately, when the tip was cleaned up, the population disappeared.

Numbers had been declining in the wild for several years before the Eastern Barred Bandicoot Recovery Team was formed in 1989.

The team moved as many of the remaining bandicoots as they could find from Hamilton to a free-ranging enclosure at Woodlands Historic Park near Melbourne Airport. In 1991, Zoos Victoria began a captive breeding program for the species. One of Australia's longest running conservation breeding programs, it continues to provide an 'insurance' population of bandicoots that will help maintain genetic diversity – the loss of which across the remaining population is of concern for the species' future health.

Currently, three free-ranging Eastern Barred Bandicoot populations are protected by fox-proof fencing at Hamilton Community Parklands, Mt Rothwell Biodiversity Interpretation Centre and Woodlands Historic Park. Some bandicoots have been relocated to fox-free Churchill Island and Phillip Island. There will hopefully be relocations to more islands off the Victorian coast in the future.

The efforts of the recovery team mean that the species' future now looks positive. Proven and new ways to protect the species are being trialled. These involve setting up large, fenced areas of native grasslands to keep foxes out, and introducing the bandicoots to isolated, predator-free islands. If successful, this could provide significant new areas of habitat for large populations. A Maremma guardian dog trial is also aiming to see if these specialised dogs can protect the bandicoots from foxes in the wild.

The Eastern Barred Bandicoot Recovery Team is made up of people from Conservation Volunteers Australia; Victoria's Department of Environment, Land, Water and Planning; Mt Rothwell Biodiversity Interpretation Centre; National Trust of Australia; Parks Victoria; Phillip Island Nature Parks; The University of Melbourne; Tiverton Property Partners and Zoos Victoria.

With their support and yours, we can all help protect and save the Eastern Barred Bandicoot.